Pernicious Marine Life

A Guide to Venomous and Poisonous Marine Animals

Complete Edition

Michael Phife

Illustrations by Kade O'Casey and Sonay N. Baker

ISBN-13:
978-1540758309

ISBN-10:
1540758303

Table of Contents

Disclaimer   1

Cartilaginous Species   2

Bony Species   18

Invertebrate Species   54

Stationary Reef Species   84

Lethal Toxins and Their Functions   91

Misconceptions   102

Final Thoughts   106

Glossary   108

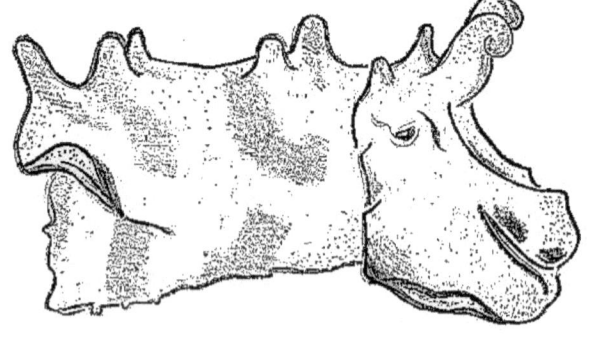

Disclaimer

The list of side-effects accompanying each species is meant to educate and represent a broad spectrum for potential reactions others have experienced in the past. These side-effects are observed through the physiological differences that each strain of venom or poison can produce on the human body, and the toxicity's impact on human health. Not every person will have the same side-effects listed per species if infliction does occur. Nor will they experience every effect. Each person will have a unique or idiosyncratic reaction, meaning that some people react very differently to certain types of venom than others. Use this guide as a reference for potential outcomes and effects as to what may happen when handling these species. Medical advice and care should always be sought if any side-effects become detrimental to one's health.

Cartilaginous Species

Stingrays

(Myliobatoidei)

Description:

This type of cartilaginous fish is also known as *Batoids*.
Commonly known species in this family include Eagle Rays
and Butterfly Rays, but the most well known would be the
Manta Ray. They are located in all Oceanic bodies, although
they prefer more tropical temperatures. Some species can
even be found in brackish water and fresh water. The size of
Stingrays varies by species. Some will stay as small as 1.3 feet
in length (including their stinger), while species such as the
Manta Ray can reach 6.5 feet and weigh in at over 3,000
pounds. They are very closely related to Sharks, despite the
differences in physical appearance. Similar to Sharks, though,
Stingrays do not have a skeletal structure. Instead, they have
hard cartilage that protects them, comparable to a form of
armor, and hard enamel for their teeth and jaw. A Stingray's
pectoral fins look very similar to birds' wings, which are
elongated and fused to their heads. Their gills are also located
on the underside of their bodies. The other noticeable
characteristic of a Ray is their large barbed stinger, which has
evolved to only be used as a self defense mechanism. Most
members of the Ray family are scavengers, feeding on small

crustaceans and fish off of the seabed. The Mobula and Manta Rays are the only exception, being filter feeders that primarily rely on plankton as their food source.

Venom:

The venom that a Stingray produces is a neurotoxin. The venom is only located in the barbs of their stingers, and they will only attack out of self defense. Stingrays almost never go out of their way to attack or act aggressively. In nearly all cases of human causalities, the Stingray was provoked, or felt provoked before striking. Unusual death or freak accidents should not be classified as an example of a Stingray's lethality. Aside from this, their venom is similar to all other forms of neurotoxin. However, their toxicity is more potent than many other sea creatures that produce neurotoxin, so caution is highly advised when handling Stingrays in the ocean. Anti-venom does exist if inflicted with a Stingray's neurotoxin. Doctors may have to remove the barbs from one's skin if punctured by them as well.

Side-Effects:

Initial Reactions:

- Swelling
- Bruising
- Numbness
- Nausea
- Diarrhea
- Headaches

Prolonged Reactions:

- Fainting
- Decreased blood pressure
- Abnormal heart rhythms
- Tremors

Severe Reactions or Reactions via Infection:

- Respiratory failure
- Seizures
- Paralysis
- Death

Skates

(Rajidae)

Description:

These are another type of cartilaginous fish that are in the
same order as Rays. There are approximately 200 species in
this family, though more are still being discovered and
diversified from Rays. One of the most popular species of
Skate would be the Guitarfish. Skates look almost identical to
that of a Stingray, except their pectoral fins make more of an
oval pattern along their bodies. Skates are generally smaller
than their counterparts, and their fins are further away from
their head. These fish are found in the Pacific, Indian, and
Atlantic Oceans, although they prefer tropical temperatures.
The size of a Skate varies on the species. The smallest
recorded species is 2.6 feet in length, while the largest species
reaches 8.2 feet. Just like Rays, the Skate is also a scavenger
that will scavenge for any small creatures it can find on the
sea floor.

Venom:

The venom that a Skate produces is a neurotoxin. However, only those with spines or barbs on their tails are venomous. Similar to that of a Stingray, the venom is only located in the barbs of the stinger. Skates will only strike out of self defense or if provoked, and have never been recorded to attack out of aggression. Humans are usually inflicted with a Skate's venom by accidentally stepping on them in sand beds. Since they burrow, this is a common occurrence to happen for people walking in the sand while at the beach. Their venom is not as a strong as a Stingray's, although medical attention is still recommended if inflicted. Anti-venom does exist for a Skate's neurotoxin.

Side-Effects:

Initial Reactions:

- Swelling
- Bruising
- Numbness
- Nausea
- Diarrhea
- Headaches

Prolonged Reactions:

- Fainting
- Decreased blood pressure
- Abnormal heart rhythms

Severe Reactions or Reactions via Infection:

- Seizures
- Paralysis
- Death

Dogfish Sharks

(Squalidae)

Description:

Dogfish Sharks are otherwise known as *Mud Sharks* or *Spiny Dogfish*. There are 119 known species within this genus. Dogfish Sharks are found in all Oceanic bodies, and prefer tropical temperatures. These cartilaginous fish vary in size by species. Some will only reach 7 inches in length, such as the Dwarf Lantern Shark, which is the smallest Shark in the world, while others will grow a little over 24 feet. They are mostly a grayish color with some hints of white along their bodies. One distinguishable characteristic about this particular species is that they do not have an anal fin. These types of Sharks feed off the bottom of the ocean instead of freely swimming for their prey. They feed on various fish and invertebrates, much like any other member of the Shark family. Dogfish have been reported to live up to, or exceed 100 years of age. This is much longer than other reported cartilaginous species, and may even be a record for fish longevity. Due to overfishing, Dogfish Sharks are now considered critically endangered worldwide and there are tight fishing restrictions on their capture.

Venom:

The venom produced by the Dogfish Shark is believed to be a neurotoxin. The venom is located in the spines of their dorsal fins. If captured, these Sharks will try to whip themselves around in order to envenom their targets for defense. As the only venomous Shark species, it may be an indication that they are directly related to Chimaeras whenever they evolved in the Devonian Era. Anti-venom does exist if inflicted by a Dogfish's venom. However, not much is known about the true effects of their venom, and reported symptoms have remained relatively mild in human afflictions.

Side-Effects:

Initial Reactions:

- Swelling
- Bruising
- Nausea
- Diarrhea
- Headaches
- Severe pain

Prolonged Reactions:

- Decreased blood pressure
- Abnormal heart rhythms

Greenland Sharks

(Somniosus Microcephalus)

Description:

The Greenland Shark is one of few Shark species that can live in the frigid waters of the Arctic Ocean. They are primarily found around Greenland, Iceland, parts of Northeast Canada and Northern Europe. Their size is comparable to a Great White's, ranging between 8 to 16 feet in length. These Sharks are either gray or a dark brown in color with spotting along their bodies. They feed on various fish and invertebrates that are located in the colder depths of the ocean. They have also been reported to be one of the longest living Sharks known to science. Some of which have reached a lifespan of over 500 years.

Poison:

The Greenland Shark's meat is poisonous to humans. It contains a chemical called *Trimethylamine N-oxide*, which is found in a few fish to help with protein stabilization. Their meat is considered a delicacy in Iceland, where it must be fermented and air dried for several months in order to remove the toxins. The side-effect of eating Greenland Shark meat before fermentation is similar to being severely intoxicated. These effects will typically wear off over the course of a day.

Side-Effects:

Initial Reactions:

- Nausea
- Headaches
- Impaired motor skills
- Impaired or slurred speech
- Temporary loss of rational/cognitive thought

Chimaera

(Chimaeridae)

Description:

Chimaeras are also known as *Rat Fish, Spookfish,* or *Ghost Sharks.* They are a type of cartilaginous fish related to both Sharks and Rays. There are 50 discovered species in this family, although there could be many that are currently undiscovered. Most of these species are located in the Abyssal Zone of the ocean, which make them hard to reach without the use of a submarine. These fish are found in all Oceanic bodies, and prefer tropical temperatures. Chimaeras broke away from other cartilaginous fish around the Devonian Era and have remained relatively unchanged ever since. The size of a Chimera varies by species, with some reaching a maximum length of 3 feet, while others can grow up to 5 feet. Their coloration is very similar to a Shark's, with the primary colors of their bodies being a gray or white. These fish are carnivores, preying on any fish or crustacean they can find.

Venom:

The venom that a Chimaera produces is believed to be a neurotoxin. This venom is located in their dorsal fins and has a very similar toxicity effect to that of a Stingray or Dogfish. Anti-venom does exist if inflicted by a Chimaeras' venom. However, the only species of a Chimaera that a human would likely ever encounter while diving would be the Ratfish. This is due to the Ratfish living in more shallow water compared to the other species. Very few afflictions have ever been reported, and those which have been envenomed by a Chimaera stated that the effects do not cause serious harm to humans. Therefore, the side-effects designated for the Chimaera species are extrapolated of what can happen, based off of what we know about Ratfish.

Side-Effects:

Initial Reactions:

- Swelling
- Numbness
- Nausea
- Diarrhea
- Headaches

Prolonged Reactions:

- Fainting
- Decreased blood pressure
- Abnormal heart rhythms

Severe Reactions:

- Tremors
- Paralysis

Bony Species

Lionfish

(Pterois)

Description:

The Lionfish is also known as the *Turkeyfish* or *Firefish*. There
are 12 species within this family, with the most notorious
being the Volitan (or Red) Lionfish, and the Devil Firefish.
These fish primarily live in the Indo-Pacific Ocean. However,
there has been a fairly recent outbreak (starting in 2003) of
Lionfish in the Atlantic Ocean. These fish are now
considered an invasive species in this region. They are killing
the sea life around the Caribbean islands, Florida Coast, and
Mediterranean Sea. The coloration of Lionfish differs by
species, but most species have zebra-liked stripes running
across their bodies. Many of these fish can be found for sale
in aquatic shops for aquariums. They will feed on a variety of
foods such as smaller fish, crustaceans, coral, and have even
been known to be cannibalistic. One major distinction
between Lionfish and their relatives of the Scorpionfish and
Stonefish family is that they will hunt for their food instead of
ambushing. These fish also swim freely and do not have low
buoyancy like Scorpionfish or Stonefish. The size of a
Lionfish depends on the species, with some only reaching a
maximum of 4 inches in length while others can grow up to

19

16 inches. The most pronounced distinction on a Lionfish would be their dorsal fin spikes.

Venom:

The venom that a Lionfish produces is a neurotoxin. The venom is located in their dorsal and pectoral fin spikes. Since they are now becoming invasive in the Atlantic Ocean, more reports of causalities resulting from humans being pricked by Lionfish spikes has increased exponentially. These inflictions can lead to a range of severe side-effects, and are even reported be lethal for people who have allergic reactions to their venom. As a result of their invasion in the Atlantic, many divers and fishermen are collecting Lionfish to be eaten at local restaurants. This practice has been put in place in the hopes of reducing fatalities, and to also cull the local sea life from being devastated further. Anti-venom does exist for a Lionfish's neurotoxin, and it is highly recommended to seek medical assistance if envenomed.

Side-Effects:

Initial Reactions:

- Swelling
- Nausea
- Diarrhea
- Severe pain
- Numbness
- Fever
- Dizziness
- Headaches
- Tingling sensations
- Chest pains

Prolonged Reactions:

- Muscle spasms
- Fainting
- Decreased blood pressure
- Abnormal heart rhythms

Severe Reactions:

- Difficulty breathing
- Convulsions
- Seizures
- Paralysis
- Shock
- Anaphylaxis
- Heart failure
- Death

Scorpionfish & Goblinfish

(Scorpaenidae)

Description:

Scorpionfish are otherwise known as *Stingfish, Firefish,* or *Dragonfish.* Goblinfish are also known as *Waspfish* or *Demon Stingers.* These 2 species of fish have been paired in the same category because they are very closely related to one another. Lionfish also belong to this family, but they have been listed separately for various reasons provided on their page. Scorpionfish and Goblinfish are notorious for envenoming humans more than most other bony fish. Most of these species of fish do not exceed a length of 10 inches. The majority of these fish have colorations and markings to camouflage against sand or rocks. Due to this trait and also their slow paced swimming, these fish like to ambush their prey. Their camouflage makes them much more of a hazard for people who snorkel and dive along the seabed.

Venom:

The venom that Scorpionfish and Goblinfish produce is a neurotoxin. The venom is located in each of their pelvic, anal, and dorsal spines. Due to their toxicity, there are no known predators to these types of fish, making them an apex predator. Some victims have even reported the pain to be so severe that they felt as if they were going to die. If someone is envenomed by one of these fish and cannot make it to a hospital right away, then applying heat to the infliction helps relieve some of the pain. Anti-venom does exist for these fish's neurotoxin.

Side-Effects:

Initial Reactions:

- Swelling
- Nausea
- Diarrhea
- Severe pain
- Numbness
- Fever
- Headaches

Prolonged Reactions:

- Muscle spasms
- Fainting
- Decreased blood pressure
- Abnormal heart rhythms

Severe Reactions:

- Difficulty breathing
- Tremors
- Paralysis
- Death

Stonefish

(Synanceia)

Description:

The Stonefish is considered one of the most lethal bony fish in the entire sea. These fish are primarily found in the Indo-Pacific Ocean, where they will camouflage themselves against rocks or the sea bed to ambush their prey. Most species are found near the coastal regions of the Australian continent. Stonefish come in a variety of true colors, but most species will be brown, orange, or red before blending into their environment. These fish can range between 1.2 and 1.8 feet long. The lethality of a Stonefish's venom sets them at the top of their food chain, making them an apex predator.

Venom:

The venom that a Stonefish produces is a neurotoxin. The venom is located in their dorsal fin spines. These fish carry one of the most potent forms of neurotoxin that exist naturally, and there have been many reported fatalities from accidentally stepping on them. Unlike many other venomous fish, the Stonefish will continue injecting toxin into its victim proportional to how much pressure is being applied to the dorsal spine itself. Therefore, the harder a person steps on the spine, the more lethal the strain of venom becomes. Many people who are inflicted by Stonefish spines end up having severe side-effects and even risk the chance of dying if not treated right away. Anti-venom does exist for a Stonefish's venom. However, many people who are inflicted may not have enough time to get to a hospital depending on the severity of the sting.

Side-Effects:

Initial Reactions:

- Swelling
- Nausea
- Diarrhea
- Severe pain
- Numbness
- Fever
- Delirium
- Headaches

Prolonged Reactions:

- Muscle spasms
- Fainting
- Decreased blood pressure
- Abnormal heart rhythms

Severe Reactions:

- Difficulty breathing
- Tremors
- Seizures
- Paralysis
- Shock
- Death

Sea Catfish

(Ariidae)

Description:

Sea Catfish are otherwise known as *Ariid Catfish*. They are related to freshwater and brackish Catfish, and share many traits with them. There are roughly 143 species of Sea Catfish, many of which are euryhaline, can live in both fresh and saltwater. They are found in all Oceanic bodies, and prefer tropical temperatures. Size varies by species, with some only reaching 8 inches in length while others can grow up to 1.7 feet. Their coloration also varies, but most are a gray complexion with hints of green and white. Sea Catfish mostly prey on crustaceans and other fish, with some species even feeding on seaweed, Echinoderms, and Cnidarians. They are commonly caught by fishermen, and are noted to be a nuisance fish when attempting to fish for larger game. Like many other Catfish, these also have barbells protruding from the lower portion of their mouths to help find food on the ocean floor.

Venom:

The venom that some species of Sea Catfish produce is called *Crinotoxin*. This same toxin can also be found in various freshwater species of Catfish as well. The venom is found in the spines of their dorsal and pectoral fins. A few species can even release this toxin from their skin upon handling them. Since Catfish live in murky regions, there is a greater risk of infection if envenomed. It is important to wash and sterilize the afflicted area immediately after contact is made to prevent such an infection. Most of the time, the venom will dissipate on its own. However, there have been severe reactions to this venom and medical assistance is required if symptoms worsen within the initial hour of infliction. Doctors may also have to remove the spine that gets left behind in the skin if inflicted.

Side-Effects:

Initial Reactions:

- Swelling
- Severe pain

<u>*Prolonged Reactions or Reactions via Infection:*</u>

- Cyanosis

- Gangrene

- Necrosis of skin

Weeverfish

(Trachinidae)

Description:

Weeverfish are an unusual type of bony fish that consists of 9 different species. Unlike most fish, Weeverfish do not have a swim bladder. This makes it possible for them to sink to the bottom of the ocean whenever they stop swimming. Taking advantage of this specialized characteristic, they will bury themselves in the sand and ambush any prey that comes near them. These fish are found in the Atlantic Ocean, although one species in this family can be found around the coast of Chile in the Pacific Ocean. Weeverfish are mostly brown or green in color, with a white or beige underside. The size varies depending on the species. Some will reach a length of only 6 inches, while others can grow up to 1.9 feet.

Venom:

The venom that a Weeverfish produces is called *Dracotoxin*. The venom is located in their dorsal fin and gill covers, which many people accidentally step on when walking over sand beds in shallow water. The spines from their dorsal fins can break off into the skin, which will then discharge venom. Pain is typically noticed after the first few minutes of being envenomed. Anti-venom does not exist for a Weeverfish's venom, yet there are other ways to treat Weeverfish stings. Dracotoxin is heat labile; therefore it can be eradicated by applying hot water to the area of infliction. Anti-histamines will also help reduce the swelling caused by the sting.

Side-Effects:

Initial Reactions:

- Swelling
- Itching
- Nausea
- Numbness
- Muscle aches
- Headaches
- Cramping
- Dizziness

Prolonged Reactions:

- Severe pain
- Fainting
- Decreased blood pressure
- Abnormal heart rhythms

Severe Reactions or Reactions via Infection:

- Difficulty breathing
- Gangrene
- Tremors
- Seizures
- Nerve damage

Toadfish

(Batrachoididae)

Description:

Toadfish are otherwise known as *Frogfish*. They are given this name because they use their swim bladders to making a foghorn-like sound to attract potential mates. There are 80 known species in this family. They live in all Oceanic bodies, with some even living in brackish water. Size varies by the species, with some only reaching 3 inches in length while others can grow up to nearly 2 feet. The coloration also varies by species, but most are a dull brown, green, or gray complexion. One unique trait these fish possess is that they are able to tolerate polluted water, and can even survive out of water for an extended period of time. Toadfish will primarily feed on smaller fish and invertebrates. They have been known to bite humans, which can lead to infection if the skin is penetrated.

Venom:

The venom that some Toadfish can produce is believed to be a neurotoxin. The venom is located in their dorsal, pectoral, and anal fin spines, as well as their gill covers. There are only a few species within this family that are venomous. Inflictions generally happen whenever these fish are stepped on while dwelling in sea beds. Anti-venom does exist if inflicted by a Toadfish's venom. The symptoms of their toxin are similar to that of a Lionfish or Stonefish, although less severe in most cases.

Side-Effects:

Initial Reactions:

- Swelling
- Nausea
- Diarrhea
- Severe pain
- Numbness
- Fever
- Headaches

- Muscle spasms
- Fainting
- Decreased blood pressure
- Abnormal heart rhythms

Severe Reactions:

- Difficulty breathing
- Tremors
- Seizures
- Paralysis
- Death

Stargazers

(Uranoscopidae)

Description:

The Stargazer is a very unique fish that has eyes fixated on the top of their heads rather than the sides. They can typically be found laying flat on the ocean floor staring upwards, thus giving them their names. There are currently 50 known species in this family, and are located in all Oceanic bodies. Size varies by species, where the smallest recorded species is only 6 inches in length, while the largest species reaches 3 feet. Stargazers are able to bury themselves underneath the sand and have colorations to blend in with the sand and rocks. Similar to that of an Anglerfish, some Stargazers have a type of lure connected to their mouths, which draw in crustaceans or smaller fish for them to consume. These attributes make them very successful at ambushing their prey.

Venom:

The venom that Stargazers produce is a neurotoxin. The venom is located on their gill covers which are called the *operculum*. The chances of being envenomed by these fish is fairly rare due to this specific location, however it can happen if they are stepped on. There are also a couple species that can produce electricity, which happens around the eye muscles of the fish. This is a unique and rare trait, considering that only a few oceanic fish have an electric organ. Anti-venom does exist if inflicted by a Stargazer's venom, though the electric shock emitted by some of the species will have to wear off on its own.

Side-Effects:

Initial Reactions:

- Swelling
- Numbness
- Nausea
- Diarrhea
- Headaches

Prolonged Reactions:

- Decreased blood pressure
- Abnormal heart rhythms
- Respiratory failure

Severe Reactions:

- Paralysis
- Death

Rabbitfish

(Siganidae)

Description:

Another name for this family of fish would be the *Spinefoot*.
The Rabbitfish family is comprised of 29 different species.
The most notorious out of these species is the *Foxface*
Rabbitfish, which often times are sold for aquariums. They are
given the name "Rabbit Fish" due to their mouth looking
similar to that of a rabbit's, along with their large, bead-like
eyes. They are found around coral reef ecosystems in the
Indo-Pacific Ocean. Rabbitfish are usually seen grazing on
algae, seaweed, or sometimes even coral. This is commonly
seen in aquarium settings where they will pick at coral if they
feel that their daily feedings are inadequate or subpar. The
size of a Rabbitfish varies depending on species. Some will
stay around 7 inches in length, while others can grow to a
maximum of 21 inches. Rabbitfish are curious in nature, and
have even been deemed as opportunistic scavengers.

Venom:

The venom that Rabbitfish produce is called *Enterotoxin*. Many aquarium enthusiasts have experienced this type of venom while working with Rabbitfish. Since they are agile swimmers and fairly curious, it is not uncommon for one to swim by someone's hand and prick it with its dorsal fin, envenoming the inflicted area. Luckily, the venom almost never causes serious harm to humans. The venom will break down the proteins within the victim's intestines, usually causing abdominal pain. The afflicted area will swell up and become purple, similar to that of a bruise. As the venom compound is broken down by heat, the venom can be treated by applying a heated substance to the location of the infliction. In most cases, applying a rag soaked in warm water will relieve the swelling after several minutes.

Side-Effects:

Initial Reactions:

- Nausea
- Diarrhea
- Bruising
- Swelling
- Stinging sensation
- Abdominal cramps
- Severe pain

Boxfish

(Ostraciidae)

Description:

The Boxfish are closely related to the family of Pufferfish. They are given this name due to their box-like or hexagonal bodies, and are relatively slow swimmers. There are 23 species within this family. These fish are found in the Indo-Pacific and Atlantic Oceans, and prefer tropical temperatures. Size varies by species, with some reaching 5 inches in length while others can grow up to 1.8 feet. Their coloration varies by species, with most being a yellow, brown, or beige. Most will also have some type of polka dot pattern along their bodies. Some Boxfish will appear to have horns protruding from their heads, which are in the genus commonly known as Cowfish. Boxfish have a small set of teeth, which they use to eat small crustaceans and mollusks. Whenever this fish is disturbed, they may remain completely motionless in hopes that their aggressors pass them by. They are also able to release a poison from their skin which is incredibly potent, even at a low concentration of 10 parts per million.

Poison:

The poison that Boxfish produce is called *Pahutoxin,* or formally known as *Ostracitoxin.* This is a form of neurotoxin which is secreted from the fish's skin. It will come out as a soapy residue and can even have a red or green coloration to it. This toxin can only be produced when the Boxfish is alive. Despite this, the residue of their poison can still be found on their skin after death, making them unsafe to eat. The only way that humans can come into contact with this toxin is if they consume or inhale it. There have been few reported effects of Boxfish toxin on humans. This is due to the poison diluting immediately upon entering open water. However, those who have had contact with *Pahutoxin* have reported similar side-effects to that of a Pufferfish's *Tetrodotoxin.*

Side-Effects:

Initial Reactions:

- Numbness
- Headaches
- Tingling sensations
- Severe muscle pains

Prolonged Reactions:

- Motor dysfunction
- Black discharge in urine
- Decreased blood pressure
- Abnormal heart rhythms
- Respiratory failure

Severe Reactions:

- Paralysis
- Asphyxiation
- Death

Pufferfish & Porcupinefish

(Tetraodontidae & Diodontidae)

Description:

Pufferfish are also known as *Blowfish* or *Balloon fish*. There are
120 known species of Pufferfish and 18 species of
Porcupinefish. These fish get their name because they will fill
their bodies with oxygen or water when threatened or
stressed, causing their spines to protrude outwards. This can
also be potentially fatal to the fish if they are experiencing
consistent stressors. Porcupinefish and Pufferfish are found
in the Indo-Pacific and Atlantic Oceans, and prefer tropical
temperatures. There are also a few species that can be found
in brackish and freshwater. Some species will only reach 3
inches in length while others can grow over 3 feet. Their
coloration varies by species, but most will have a brown,
white, or black contrast to them. These fish have fused teeth
to resemble a beak, which they use to crack open the
exoskeletons of crustaceans and mollusk shells, their main
food source. One common misconception about these fish is
that people think their spines are venomous, which is not
true. Their meat is the only toxic part of their bodies.

Poison:

The poison that some Pufferfish and Porcupinefish produce is called *Tetrodotoxin*. This is a form of neurotoxin, and is about 12 times more deadly than cyanide. This toxin is found in the liver, skin, intestines and reproductive organs of the fish, and can only affect humans if their meat is ingested. Japan is one of the most notorious countries to consume this fish, which they call *fugu*. If prepared incorrectly, it can be fatal. The effects can be experienced anywhere from 10 minutes to 6 hours after consumption. Death can occur between or after this timeframe as well. Those who have not died from eating toxic species of Pufferfish or Porcupinefish have reported to be in a near-death state for several days afterwards.

Side-Effects:

Initial Reactions:

- Nausea
- Diarrhea
- Numbness
- Headaches
- Dizziness
- Dilated pupils
- Tingling sensations
- Motor dysfunction

Prolonged Reactions:

- Decreased blood pressure
- Abnormal heart rhythms
- Respiratory failure

Severe Reactions:

- Paralysis
- Seizures
- Asphyxiation
- Coma
- Death

Sea Snakes

(Hydrophiinae)

Description:

This family of snakes is also known as *Coral Reef Snakes* or *Sea Serpents*. There are currently 62 known species in this particular family, with each species being venomous. Their relatives include cobras, adders, mambas, and copperheads to name a few. Sea Snakes are found throughout coastal areas of the Pacific and Indian Ocean. They need to stay closer to shore due to the fact that they do not have gills and must resurface periodically to breathe. Their appearance resembles that of an eel, although much more slender in shape. The Sea Snake has a range of colors depending on the species. Some will be a solid color such as a greenish-yellow or brown, while others may have stripes along their body. Their traits are the same as that of their terrestrial relatives, yet their sensory abilities may be distorted when submerged in water. They also have salt glands below their tongue, which releases salt out of their bodies whenever their tongue is flicked. It is believed that Sea Snakes may have a separate set of receptors all together for underwater survival, although it is not entirely clear. Their size varies based on species, with the smallest ranging close to 4 feet and the largest ranging to nearly 10 feet in length.

Venom:

The venom that Sea Snakes produce is a neurotoxin, yet some species have also been found to have their venom mixed with *Cardiotoxin* and *Cytotoxin*. Neurotoxin is one of the most common types of toxins found in nature and can lead to a plethora of different side effects, depending on the species. Sea Snakes are considered highly venomous to humans, yet there is only about a 20% to 33% chance a person will become injected with their venom if bitten. This is due to most of their bites being blank, where they are not able to excrete enough venom from their glands in time whenever provoked. However, bites are usually never felt and there is almost no prior indication that someone has been bitten. If the bite is venomous, the symptoms will occur anywhere from 30 minutes to 1-2 hours later. Luckily, anti-venom does exist for these bites if a person is afflicted.

Side-Effects:

Initial Reactions:

- Headaches
- Sweating
- Dehydration
- Diarrhea
- Nausea

Prolonged Reactions:

- Swelling of tongue
- Stiffness or tenderness of muscles

Severe Reactions:

- Breakdown of skeletal muscle tissue
- Paralysis
- Death

Invertebrate Species

Blue-Ringed Octopus

(Hapalochlaena)

Description:

The Blue-Ringed Octopus is the one of the only members of the Cephalopod family that produces a sufficient amount of venom to harm humans. Although, it is worth noting that all members of the Cephalopod family are venomous to a degree. However, other member's toxins can only be felt by smaller animals such as fish or crustaceans. They are commonly found within shallow areas of water, such as a coral reef system or a tide pool. Their range consists of the Pacific and Indo-Pacific Ocean, but they are primarily located around Australia and parts of Japan. The Blue-Ringed Octopus ranks as the most venomous marine animal, and is considered one of the most venomous animals on the planet. They are easily identifiable by their yellowish-beige complexion and blue rings along their body. This octopus is rather small in size, growing no longer than 8 inches, but typically staying around 5 to 6 inches from their head to the base of their tentacles as adults.

Venom:

The venom secreted by this octopus is called *Maculotoxin*, which is a form of *Tetrodotoxin* under the classification of neurotoxins. This is one of the most powerful natural strains of venom in the world. The Blue-Ringed Octopus will create this venom with their mucus, but it can also be produced through their skin after prolonged contact. This Cephalopod will only bite if provoked or startled. Any type of bite by this octopus will lead to paralysis, in milder cases, but often times will cause death. On average, humans will only have a few minutes before they feel the side-effects of the venom. There is currently no anti-venom available if afflicted; only ventilation machines to help stabilize breathing in case of respiratory failure.

Side-Effects:

Initial Reactions:

- Nausea
- Numbness

Severe Reactions:

- Temporary Blindness
- Paralysis
- Respiratory failure
- Heart failure
- Death

Flamboyant Cuttlefish

(Metasepia Pfefferi)

Description:

The Flamboyant Cuttlefish is the only member of Cuttlefish that is venomous enough to harm humans. They are found in the Indo-Pacific Ocean primarily around Australia, Malaysia, Philippines, Indonesia and New Guinea. They grow to 2-3 inches in length, making them one of the smaller species of Cuttlefish. They are naturally dark brown but consistently change colors via camouflage. Unlike many members in this family, the Flamboyant Cuttlefish is able to go through a series of complex colors through camouflage which is used to disorient and mesmerize its prey. This can be seen as a strobe effect and is sometimes hypnotic even to humans. They are also able to walk on the sea floor with their tentacles and mantle, which tricks their prey into thinking it is a different animal altogether. Just like other members of the Cephalopod family, the Flamboyant Cuttlefish has a life expectancy between 2 to 5 years, with most reaching their longevity at the age of 3.

Venom:

The venom that the Flamboyant Cuttlefish produces is currently unknown. By speculation and with their relation to the Blue-Ringed Octopus, it is presumed that their venom is *Maculotoxin* due to the similar effects and the potency it gives off. The toxin is located in their muscle tissue which can pass on to its victim after prolonged touch, although being inflicted with a Flamboyant Cuttlefish's venom is rare since they tend to avoid humans. Many who are envenomed by this species typically handle them unintentionally in the wild.

Side-Effects:

Initial Reactions:

- Nausea
- Numbness

Severe Reactions:

- Paralysis
- Respiratory failure
- Heart failure
- Death

Cone Snails

(Conus)

Description:

The Cone Snail is also called the *Cigarette Snail*, because a person has just enough time to finish smoking one cigarette before dying from their venom. Cone Snails are one of the only discovered Gastropods that are venomous. There are over 800 reported species of Cone Snails throughout all Oceanic bodies. These types of Snails favor tropical waters, and most are found in the Indo-Pacific Ocean. Most species will stay around 4-6 inches in length, while the larger species can grow to 9 inches. Just like all Gastropods, Cone Snails are carnivorous scavengers and will eat anything they can find, even other Snails. The shells of these Snails are very colorful and have vibrant markings. They can come in many colors, and each species has a different pattern along the shell itself. There has been a huge market on these shells over the course of human history, which people have collected to make jewelry out of.

Venom:

The venom that Cone Snails produce is called *Conotoxin*, which is a form of neurotoxin. Many of the smaller species will only have effects similar to that of a bee sting. However, some of the larger species such as the Textile Cone Snail, which is also the most venomous, have enough venom to kill humans. The way in which this Snail envenoms its prey is with a harpoon-like tooth. They will fire venom at their target, which can be shot in any direction they desire. Some of the larger Cone Snails can shoot their harpoons with such force that it will break through a diving suit and penetrate the skin. The larger Cone Snails are so potent with their venom that even a single drop can kill a dozen or so humans. There are currently strains of their venom being used in the production of pain killers, and have had more positive effects on patients than the use of morphine. Anti-venom does not currently exist for a Cone Snail's toxins. If envenomed, there may not even be enough time to make it to a hospital if inflicted by larger species in this family.

Side-Effects:

Initial Reactions:

- Swelling
- Numbness
- Nausea
- Headaches

Prolonged Reactions:

- Fainting
- Decreased blood pressure
- Abnormal heart rhythms
- Tremors
- Respiratory failure

Severe Reactions:

- Shutdown of nervous system
- Seizures
- Paralysis
- Death

Sea Stars

(Asteroidea)

Description:

Another common name for these creatures is the *Starfish*.
This marine animal has stood the test of time, and has been
around for over 450 million years with almost no
evolutionary changes. There are over 2,000 species of Sea
Stars that are known to exist. They are found in all Oceanic
bodies, and can tolerate most temperatures. Sea Stars come in
a variety of colors, and some are even prized as ornaments
for their coloration. One commonly known fact about Sea
Stars is that they have the ability to regenerate parts of their
body that have been damaged. As long as the brain or middle
section of the Sea Star remains intact, they can typically
regenerate most wounds or lost arms. Depending on the
species, some Sea Stars will have 5 arms while others can
have up to 40. Unlike most animal life forms, Sea Stars do
not have a central nervous system. Instead, they have
something called net nerves, which tell the Sea Star how to
respond, how to move and how to function. It is believed
that many oceanic creatures evolved from the Sea Star, and
later went on to establish a central nervous system after
branching off. Sea Stars can also reproduce both sexually and

asexually, depending on the species. The lifespan of a Sea Star is completely dependent on the species and natural conditions, but some have been reported to live over 100 years. Out of all of the discovered species, only a handful of Sea Stars are venomous.

Venom:

The venom that certain species of Sea Stars produce is called *Saponin*. This venom is released from their tissue or spines. Most of the time, the spines will break off into the skin and the toxicity effects will last for several hours. The most popular venomous Sea Stars would be the Crown of Thorns, Sun Star and Leather Star. These 3 species are generally the only ones to have ill effects on humans after exposure to their toxin. The potency of these Sea Stars varies, but typically the venom is not lethal. Anti-venom does exist for *Saponins*, but many people will wait for the effects to wear off since they are almost never severe. However, the spines may have to be removed by doctors and treatment should be sought if side-effects do not dissipate after several hours.

Side-Effects:

Initial Reactions:

- Nausea
- Skin Irritation
- Swelling
- Burning sensation
- Persistent bleeding

Sea Urchins

(Echinoidea)

Description:

The family of Sea Urchins is a diverse species in regards to oceanic life. There are over 950 known species of Urchins, and they are located in every Oceanic body. The most commonly known species would be the Long Spine Urchin, Pencil Urchin, and Tuxedo Urchin, which can all be found in aquatic shops. These creatures are related to Sea Stars and Sea Cucumbers, all of which are under the classification *Echinoderms*. Urchins can come in almost every natural color such as red, blue, purple, brown, green, black, yellow, and so forth. Most Urchins are in the shape of a ball or oval and will have spines protruding out of them. They are considered opportunistic feeders and will consume anything that they can locate on the seabed. Although, some species are known to eat smaller fish and even other Urchins. These creatures are slow moving, and many will even appear to be stationary at first glance. Many people confuse Urchins as forms of coral due to this.

Venom:

The venom that Sea Urchins can produce is a neurotoxin. However, it is important to note that not all Urchins are venomous. Some of the species that are reported to be venomous would include the Long Spine Urchin, Fire Urchin, Purple Sea Urchins, Radiant Sea Urchins and the Flower Urchin (which is the most potent). Unlike the other species of Urchins, the Flower Urchin has both a strain of neurotoxin and also *Peditoxin*. These two forms of toxicity lead to a very lethal sting, and will even cause convulsions or shock upon infliction. If a person is pricked by a Sea Urchin the spine will typically break off into the person's skin. There are not a lot of easy ways to get these out, and the spine will continue pumping toxins into the skin until the venom dissipates from the spine. Anti-venom does exist if inflicted with a Sea Urchin's venom. Many people recover from the effects after several hours, although medical attention is recommended if inflicted with a Flower Urchin's venom.

Side-Effects:

Initial Reactions:

- Swelling
- Numbness
- Nausea
- Diarrhea
- Headaches

Prolonged Reactions:

- Difficulty speaking
- Muscle relaxation
- Fainting
- Decreased blood pressure
- Abnormal heart rhythms

Severe Reactions or Reactions via Infection:

- Tremors
- Respiratory failure
- Convulsions (Flower Urchin)
- Shock (Flower Urchin)
- Paralysis
- Death

Ribbon Worms

(Nemertea)

Description:

Ribbon Worms are otherwise known by their scientific name, *Nemertea*. There are roughly 1,400 species in this family. These worms are found in all Oceanic bodies, with some being found in freshwater and on land. They are located on sea beds, with many dwelling underneath rocks or in cave systems. Most of these species are vibrant colors of red, orange, purple, yellow, and green. Some Ribbon Worms will only reach 1 centimeter in length, while others can grow over 100 feet. Though, the majority will stay around 6 to 8 inches. These worms are voracious scavengers. They are able to expand their bodies to swallow prey whole. Many will feed on invertebrates and even fish, while others are solely filter feeders. They are able to catch their prey with a tubular mouth called the *proboscis*. Since Ribbon Worms lack a true head, they are able to siphon their prey into their bodies until it asphyxiates or is stabbed. Some species can inject venom into their targets with a needle-like appendage, which is located inside of their bodies. One attribute that these worms have, similar to Starfish, is that they are able to regenerate parts of their bodies.

Poison/Venom:

The poison that some species of Ribbon Worms produce is called *Tetrodotoxin*. This is the same type of poison that Pufferfish generate. This is one of the most toxic forms of poison that occurs naturally in the world. It has been rated to be over 1,200 times more potent than cyanide. The only way that humans can come into contact with this poison is if a person accidentally consumes the worm. If inflicted, immediate medical help is to be sought. There is an estimated time frame of 10 minutes to 6 hours before a human can die from their potency. Other species of Ribbon Worms can produce neurotoxin. This has similar, although inflects less extreme effects than the species that secrete *Tetrodotoxin*. The worm must be ingested or strike a person with their appendage in order to become inflicted. It is important to note that only certain species of Ribbon Worms are poisonous or venomous. These two strains of toxin have only been found in a handful of Ribbon Worms so far.

Side-Effects:

Initial Reactions:

- Nausea
- Diarrhea
- Numbness
- Headaches
- Dizziness
- Delirium
- Severe pain

Prolonged Reactions:

- Loss of sensation
- Motor dysfunction
- Fainting
- Decreased blood pressure
- Abnormal heart rhythms
- Respiratory failure

Severe Reactions:

- Paralysis
- Asphyxiation
- Death

Bearded Fireworm

(Hermodice)

Description:

The Bearded Fireworm is one of the only members of the Bristle Worm family that is venomous. This species is located in the Atlantic Ocean and typically dwell underneath rocks or in caves. They can also be found on coral reef systems closer to shore. The Bearded Fireworm can range anywhere from 6 to 12 inches in length, depending on the food sources they can acquire. These worms eat any type of decaying matter, along with coral. Their colorations can either be red, green or yellow, with a beige line running across their bodies. The bristles on their segmented plates are a white color, and can resemble certain types of caterpillars or asps.

Venom:

The venom produced by this worm is a neurotoxin. The venom is found within their bristles, which break off into the skin if they are touched. Humans are mainly inflicted by these worms if they step into sand beds or lift up rocks where they tend to hide. The symptoms of the venom are painful and will cause irritation the moment they make contact with skin. Anti-venom does exist for Bristleworm venom, and the bristles can be removed by applying tape to the inflicted area and peeling it off, similar to any other type of Bristleworm affliction.

Side-Effects:

Initial Reactions:

- Swelling
- Irritation
- Numbness
- Nausea
- Dizziness
- Severe pain

Jellyfish

(Medusozoa)

Description:

Despite how common it is to run into Jellyfish in the ocean, they are one of the least understood species of marine life due to their biological complexity. Little has changed in their evolutionary process, which dates back 500-700 million years. This also makes them older than nearly every reported species on the planet. There are roughly 200 species of Jellyfish that have been discovered so far. They are found in all Oceanic bodies, and can even tolerate unsuitable water conditions. Some species can even be found in brackish and freshwater. Most Jellyfish are translucent in color, with some species having specific patterns on their bells, such as dots or stripes. They are bioluminescent, which means they are able to glow different colors based on the proteins in their cells. Their bodies are comprised of 96% water, which is higher than any other species in a body-to-water ratio. Some Jellyfish are only 1 centimeter in length, while others can grow larger than 6 feet (including tendrils). There are even a couple species, such as the Lion's Mane Jellyfish, which have been reported to grow anywhere between 80-120 feet in length. Many Jellyfish species will form a *bloom*, which is a congregation of multiple

77

Jellyfish in one area. Jellyfish rely on their translucent tendrils to catch their prey. Their tendrils act almost like a net, and will latch onto anything that passes through them. Their diet consists of small fish, fish eggs, crustaceans, micro organisms, and even other Jellyfish. One thing that we are currently unable to determine about Jellyfish is their age. This is due to the fact that many can reproduce asexually, and can even regenerate their old cells into new ones. This is being studied by scientists to hopefully be used in medicine and cellular regeneration for humans.

Venom:

The venom that most Jellyfish produce is called *Cnidocyte*. This is found in the tendrils of the Jellyfish. The tendrils work similar to shooting off a firearm. Whenever something rubs against their tendrils, the nematocysts (explosive cells found in the tendrils) will pierce into the skin and begin injecting venom into the perpetrator. Even after a Jellyfish dies, the nematocysts are still able to be fired off. This is why there are many people who get stung by Jellyfish that are washed up on beaches. The severity of this venom is completely dependent on the species. There are a few species, such as the Box Jellyfish, that are potent enough to kill a human within minutes. Jellyfish stings are a high occurrence worldwide, and caution should be taken if there are known Jellyfish sightings in the area. There is a widely known misconception about how to treat Jellyfish stings. A number of people in the past have mentioned applying ammonia or urine onto the afflicted area. This is not only false, but can speed up the release of the venom. The best methods are to remove the nematocysts from the skin with a sharp object, and to apply vinegar or salt water onto the wound. Medical attention should be sought immediately if symptoms become intolerable or severe.

Side-Effects:

Initial Reactions:

- Nausea
- Diarrhea
- Numbness
- Fever
- Tingling sensations
- Severe stinging pain

Prolonged Reactions:

- Rash
- Welts
- Muscle spasms
- Respiratory failure

Severe Reactions:

- Asphyxiation
- Coma
- Death

Man O' War

(Physalia)

Description:

The Man O' War is also known as *Blue Bottles* or *Ocean Terrors*. There are only two types of species in this family, which are the Indo-Pacific Man O' War and the infamous Portuguese Man O' War. Their appearance is very similar to that of a Jellyfish, yet they are very different in terms of their biological structure. These species are comprised of four different polyp colonies, each playing their own role as a part of the collective organism. They are found in the Indian, Pacific and Atlantic Oceans. Their coloration is translucent, similar to that of a Jellyfish, although many will have a blue tint to them. The Indo-Pacific species will grow to around 6 inches in length for its bell, with tendrils reaching approximately 30 feet. The Portuguese Man O' War is much larger, with the bell doubling in size and the tendrils ranging in between 30 to 160 feet in length. Their diet consists of anything that will pass by their tendrils. In most cases this will include crustaceans, small fish and micro organisms such as plankton. These species have been much more abundant in sightings over the years, more so than Jellyfish in some regions.

Venom:

The venom that a Man O' War produces is called *Cnidocyte*. The toxin is found in their nematocysts, which will shoot out from the tendril and onto its victim. This is the same type of venom produced by Jellyfish with nearly identical effects. Though, the treatment for Man O' War stings is not exactly the same as Jellyfish stings. Saltwater and vinegar should still be used to help alleviate the pain from the nematocysts. However, this method should only be used if the infliction is large and severe. Using saltwater or ammonia based vinegars have been reported to sometimes speed up the venom in smaller scale cases. Shaving off areas where the sting occurred can also help remove the nematocysts. Hot compression packs should then be used to help break down the toxicity inside of the skin. Unlike Jellyfish stings, a Man O' War sting can lead to infection very easily. After being stung, observation for the next couple days is highly recommended to determine if the side effects are becoming detrimental. If so, then immediate medical help should be sought. Man O' War stings are almost never fatal, unless someone has been stung by an entire bloom.

Side-Effects:

Initial Reactions:

- Nausea

- Diarrhea

- Numbness

- Fever

- Swelling

- Tingling sensations

- Severe stinging pain

Prolonged Reactions:

- Rash

- Welts

- Muscle spasms

- Respiratory failure

Severe Reactions:

- Asphyxiation

- Shock

Stationary Reef Species

Anemones

(Actiniaria)

Description:

Anemones are another venomous species under the classification of *Cnidaria*. There are over 1,000 known species within this family and are located in all Oceanic bodies. Their coloration is dependent on the algae, Dinoflagellates or Zooxanthella that resides within their tendrils, most of the time being either red or green. Many Anemones are only toxic to fish and some invertebrates. They can blend in with Coral since most species are found within reef systems. However, unlike Coral, Anemones do not have a skeletal structure, which is the main difference between the two families. A few fish, such as members in the Damselfish family, have a resistance to Anemone stings and form symbiotic relationships with them. There are only a couple of species which are venomous to humans, which are several species of Carpet Anemones and Hell's Fire Anemones.

Venom:

The venom that Anemones produce is called *Cnidocyte*. The stings from these species are similar to that of Jellyfish, although the *Cnidocytes* in their tendrils are less potent. The side-effects associated with touching one of these Anemones is very similar to Fire Coral. The best method for treating Anemone stings is to rinse the affected area under warm water. The nematocysts are heat labile and will dissipate over time with the heat. Other recommended methods of relief are pouring salt water or vinegar over the infliction. Most cases are never severe, and the pain will alleviate over a couple of hours.

Side-Effects:

Initial Reactions:

- Rash
- Burning sensations
- Numbness
- Severe pain

Fire Coral

(Millepora)

Description:

All Corals are venomous to some degree, as they are in the *Cnidaria* family. However, most are only toxic to other Coral species. The exception to this would be Fire Coral. This species is more closely related to Jellyfish and Anemones than they are to other forms of Coral. This particular Coral leaves a nasty sting when touched by humans. This type of Coral is located in the Indian, Pacific and Atlantic Oceans. The coloration of this species is usually yellow, orange or red.

Venom:

The venom that Fire Coral produces is called *Cnidocyte*. These types of stings are similar to that of Jellyfish stings, although the *Cnidocytes* produced by their branches are less potent. The nematocysts are located on the polyps of the Coral itself, rather than tendrils which Jellyfish and Man O' War have. Treatment of Fire Coral stings is very similar to the treatment of Jellyfish stings. Apply vinegar or salt water to the afflicted area to wash it off. Hot water also works very well due to the nematocysts being heat labile. Medical assistance may be required if the rash or pain does not dissipate after several hours.

Side-Effects:

Initial Reactions:

- Rash
- Burning Sensations
- Numbness
- Severe pain

Sea Sponges

(Porifera)

Description:

Sea Sponges are a type of filter feeder that live on sand beds, typically found alongside Coral. There are roughly 10,000 species within this family. They are located in all Oceanic bodies, and mostly found in tropical temperatures. Just like Coral, they have a vast array of colorations. Sponges are an immobile organism and rely on their surroundings to stay alive. They are completely dependent on ocean water to help them obtain and digest food, along with absorbing oxygen. Since Sponges are rooted in place, some of them must create a type of defense mechanism to deter predators.

Venom:

The venom that some Sea Sponges produce is a *Cytotoxin* created by *Polytheonamides*. This is found in both their metabolites and *spicules*, which are spiny structural elements that cover the organism. These will drive predators away either by touch, smell or taste. Some species of Sponges are toxic to humans whenever they are handled. The *spicules* will break off into the skin and begin causing irritation. However, the potency is minor compared to most other venomous creatures. Using adhesive tape is the best way to get the *spicules* out of the skin. Aside from this, applying vinegar to the afflicted area helps alleviate some of the pain. There have been very few known cases of this ever becoming detrimental to humans. Divers and aquarium enthusiasts are encouraged to wear gloves when handling Sea Sponges.

Side-Effects:

Initial Reactions:

- Rash
- Burning sensations
- Skin irritation
- Numbness

Lethal Toxins and Their Functions

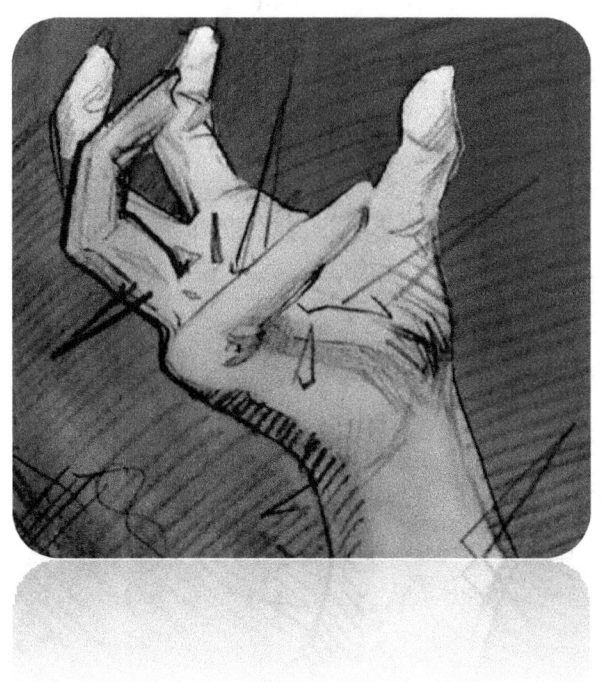

Conotoxin

$(C_{52}H_{78}N_{20}O_{15}S_4)$

As mentioned previously, this strain of toxin has proven to be lethal to humans within a matter of minutes. It is a peptide that attacks the nervous system and reconfigures the ion channels in the human body. In layman's terms, Conotoxin contains chains of specific amino acids that are injected into the blood stream by a Cone Snail. The amino acids from the venom are carried over to channels that regulate sodium and other key elements in our cells. They begin to modify the cell channels once introduced to these areas. This will shut down parts of the central nervous system, depending on where the toxin inhibits and what type of toxin has been introduced. There are five known types of Conotoxin, and each will attack a different area of the nervous system. They are categorized as followed, proceeded by which area they affect:

- Alpha – nerves and muscles
- Delta – sodium channels that depend on circulatory voltage
- Kappa – potassium channels
- Mu – sodium channels in the muscles and muscle tissue
- Omega – calcium channels that depend on circulatory voltage

Each strain of Conotoxin can be found in various species of Cone Snails. Some of the Snails will release multiple types of this toxin, while other species may only excrete a single strand.

One of the reasons that Conotoxin is not able to be treated by anti-venom is because the toxin is constantly mutating. Cone Snails have to keep adapting as fish and other species evolve to resist some of the effects of the toxin's potency. Therefore, the chemical makeup of their venom also changes, or even mutates, as the Snail adapts. This also leads Conotoxin's chemical structure to duplicate and combine with the types listed above. As an example, one species of Snail could just carry type Alpha, while another species could have a combination of Delta and Omega. This becomes problematic for scientists when trying to concoct anti-venom as combinations can alter, differ and even change from one species to the next. On top of this, Conotoxin has many other functions that are currently unknown at a molecular level.

Tetrodotoxin & Maculotoxin

$(C_{11}H_{17}N_3O_8)$

Tetrodotoxin (abbreviated as TTX) and Maculotoxin are considered the most lethal natural toxins in the world. Despite them being two different strains of toxin, they share the same chemical structure and makeup. The only difference between these two is that Tetrodotoxin is a poison while Maculotoxin is secreted as venom. Whenever these toxins are presented in the body, they inhibit excited cells (neurons that connect to different areas of the nervous system). The excited cells move through sodium channels within the membranes of the nervous system. As the toxins move through the nervous system, they will block passages that ions need to access, and will disable the function of the ions all together. This happens because the molecular shape of TTX is able to plug ion channels perfectly which prevents ion flow. This is when things become dire and a plethora of concerns for the human body arise. With the sodium channels blocked, the entirety of the nervous system will begin to shut down. It starts to affect the muscular system, followed by the endocrine, respiratory and cardiovascular systems. Depending on the amount of toxin that enters the body, the effects may either be prolonged or happen relatively quickly. Death can occur within 20 minutes to 6 hours just from a mere 0.5 milligrams of the poison or venom.

As this is one of the most deadly strains of toxin known to man, there is also no cure for it. There are only treatments to help relieve some of the side-effects, which can potentially save a person's life. This would be the use of respirators to help stabilize breathing, or applying pressure to the infliction along with a respirator if somebody has encountered a run-in with a Blue-Ringed Octopus.

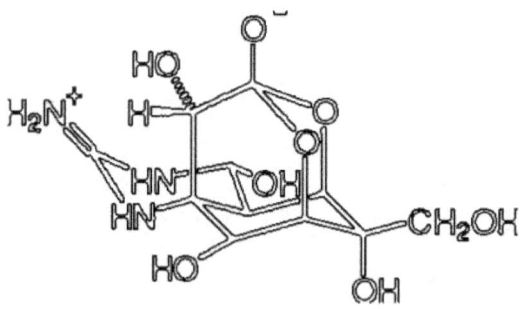

Pahutoxin or Ostracitoxin

$(C_{23}H_{46}ClNO_4)$

This toxin works similarly to Tetrodotoxin at a molecular level. Pahutoxin will bind to excited cells and enter sodium channels after being introduced into the body. This will move the toxin into the nervous system where it will begin blocking ion passages because the molecular structure fitting perfectly inside of the passages similar to a puzzle piece. This prevents ions from being entered into the nervous system, which then triggers a chain of detrimental effects on the body. Once it has made its way to the nervous system, it will start to affect the muscular, endocrine, urinary, respiratory and circulatory systems. However, one difference with this poison is that it will start to rupture red blood cells, which will cause the toxin to move throughout the blood stream at a faster rate.

Pahutoxin is not able to undergo dialysis, which means that the toxin will produce a foamy consistency when it is released in water or any other aqueous solution. This can be seen clearly if a Boxfish excretes its poison in an aquarium or a small body of water. The poison will emit as a red or green color and feel like detergent to the skin. Very few incidents have occurred with Boxfish poisoning humans, as the toxin will dissipate rather quickly if it is expelled in the ocean.

97

There is no known cure for Pahutoxin, mostly because it is highly uncommon to experience. It's also due to the molecular structure working as a nonpeptide, similar to that of TTX.

Palytoxin

$(C_{129}H_{223}N_3O_{54})$

Although this poison is not primarily created by a marine animal, its potency is lethal enough to kill humans at a rapid pace. Palytoxin is a non-protein toxin and can be hosted by various marine organisms. The chemical composition is that of a fatty alcohol, similar to ethylene and esters. One way to prove the legitimacy of this is from Palytoxin's chemical formula. The long chains of carbon will bind together, and accompany -OH chains (hydroxyl groups) which branch off and create a primary alcohol. However, this toxin behaves differently than alcohols humans typically encounter in substances or medicine. Whenever Palytoxin enters the body, it contracts blood vessels in the arteries which can cause hemorrhages. This will also reduce blood flow and turn the skin pale from the lack of body heat via blood regulation. The next phase is that the toxin will target potassium and sodium channels, which will begin killing off cells. This toxin does not confine itself to one particular area; it will affect every cell in the human body. Many scientists even theorize that Palytoxins are a culprit for Ciguatera and other food borne illnesses.

Unlike other poisons, this toxin can go through various food chains in the wild and inhibit many species. For example, Dinoflagellates are the primary source of this toxin. These protists can dwell in certain types of Coral such as Seaweed Coral (this species has been found to have the highest occurrence of Palytoxin so far). Whenever predator species consume these host species, they will in turn inhibit Palytoxin in their bodies. Though, this is not necessarily lethal for every species that ingests it. The predator species will now be a carrier, and the food chain will continue until it eventually wipes out a species that is not immune to the effects. However, the further up the food chain this toxin goes, the more persistent the toxin becomes. This means that the chances of a subsequent predator eating the prior species can be dire. Despite this, Palytoxin is lethal in nearly every stage for humans. Fortunately, scientists have been able to synthesize this toxin and concentrate it into lesser alcohols such as carboxylic (acetic) acid. The only downside to humans being poisoned by Palytoxin is that it reacts with blood vessels at a rapid pace and will cause death shortly after. This makes it difficult or even impossible to cure after exposure to the toxin. The only successful treatments to date have been on animals in laboratory studies after immediate exposure to Palytoxin.

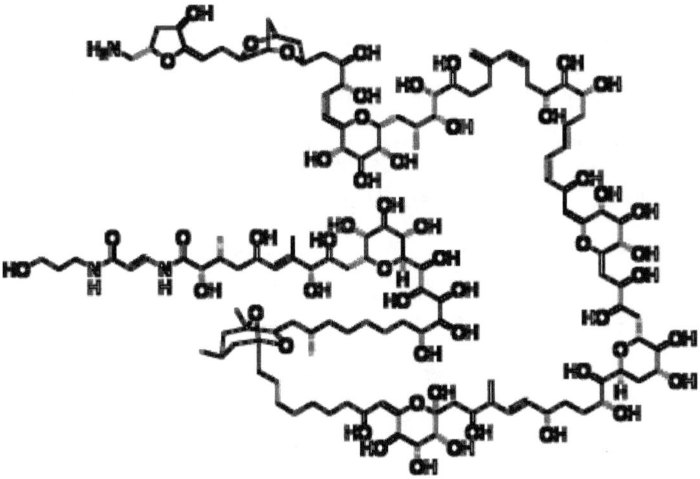

H₂N

Misconceptions

There are a few misconceptions about certain types of marine life that people believe to be venomous or poisonous. Each species will be pointed out in the following pages, along with a further explanation about them. This will also clear up why people may have thought they are or were toxic. The following species are the most noteworthy to elaborate on:

Bristle Worms & Other Annelids

(Polychaete)

Bristle Worms are a common nuisance species found in marine aquariums. They are found underneath rocks and in sand beds, and are typically active at night. Some people have the misconception that Bristle Worms, or any other type of segmented worm in this family is venomous. Upon handling them, there is a stinging or burning sensation, followed by numbness. This leads people to believe they are venomous. This is just an effect of the bristles themselves, where they will break off and stick inside of the skin. The only species of Bristle Worm that is venomous would be the Bearded Fireworm, which was mentioned previously.

Nudibranchs

(Nudibranchia)

This is a type of shell-less Gastropod that is very vibrant in color. Many will be a bright blue with stripes running along their bodies. Whenever we see vibrantly marked animals in the wild, we automatically assume that there is poison affiliated with them. This is only partially true with Nudibranchs. These snails eat certain Jellyfish, Anemones, Coral and Hydroids. Whenever ingested, the nematocysts from those species will be stored in an area of the Nudibranch's body called the *cerata*. The venom from the stored nematocysts will then act as part of a Nudibranchs' defense mechanism. This means that a Nudibranch will acquire the venom from those species and use it as its own. This will give off the same side-effects as touching a real species in the *Cnidaria* family. Caution is advised when handling these Gastropods in the wild. They may eat the notorious Box Jellyfish, and the side-effects could potentially be fatal. Some species of Nudibranchs may also emit a type of acid from their bodies. This generally only effects smaller marine species, but it may lead to slight skin irritation for humans.

Surgeonfish

(Acanthuridae)

Another name for these fish would be *Tangs* or *Spade Fish*. It was thought that their caudal (tail) fins produced venom which would harm or paralyze any predators that came near them. Although their tails are sharp and can even slice through human skin, there is no venom found within the tail itself. Scientists are still debating if there is venom found in the tails of any species of Surgeonfish.

Final Thoughts

There may have been some species that were not covered in this guide, or may have been purposely left out. This is simply because their toxicity is not potent enough to inflict much harm, if any, to humans. Or they are much more uncommon to encounter, thus making any notation about them unnecessary. One example of this would be the Sea Cucumber, where its toxins do very little to humans unless one happens to rub their eyes after handling them. There is also the failure to mention strains of toxins such as *Ciguatera, Cyanotoxin, Maitotoxin, Clupeotoxin, Saxitoxin, etc.* They have been left out due to marine animals not producing these toxins themselves. They require certain factors such as bacteria or Dinoflagellates to make the organism potent.

This guide was put together in hopes of giving respect to oceanic life and educating the reader about marine toxicology. There are far too many people who are unknowingly harmed each year by venomous sea creatures. The goal is to reduce the amount of annual causalities that are either impaired or killed by pernicious marine life.

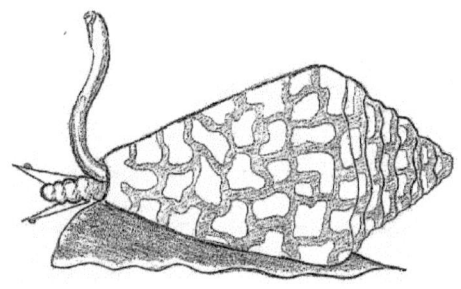

Glossary

Abyssal Zone – a pelagic zone in the ocean that is roughly 13,000-20,000 feet below the surface; never receives natural light.

Anal fin – pair of fins located underneath the tail of a fish.

Anaphylaxis – an allergic reaction to a toxin where the victim's body becomes extremely sensitive.

Annelids - segmented worms that are found both on land and at sea.

Apex predator – a predator that is at the top of its food chain.

Bioluminescence – the emission of light by any living organism.

Bloom – groups of Jellyfish or Man O' War, usually consisting of clusters that range from hundreds or thousands.

Brackish water – whenever saltwater and freshwater mix, usually happens in rivers or estuaries.

Buoyancy – the ability to float in water.

Cardiotoxin – a toxin that affects the heart; can weaken the heart and causes dysfunctions with blood flow to the aorta, along with muscular damage.

Caudal fin – otherwise known as the tail fin; used for propulsion.

Cephalopod – a class of animals which consists of Octopus, Squid, Nautilus and Cuttlefish.

Cerata – lateral horn growths on the bodies of Nudibranchs.

Ciguatera – a food borne illness caused by eating fish that is contaminated with certain types of Dinoflagellates; consumption typically results in severe nausea and even cardiac or neurological symptoms.

Cnidarians – a class of animals which consist of Jellyfish, Man O' War, Anemones, Hydras and Coral.

Cnidocyte – an explosive cell that contains toxins; found in Cnidarian species which gives off a stinging sensation and a multitude of other effects.

Clupeotoxin – a food borne illness that occurs after eating fish contaminated by a specific type of plankton; can be found in Herring, Sardines, Anchovies, Tarpons and several other species.

Conotoxin – a lethal type of neurotoxin that is produced by Cone Snails; manipulates ions in the body.

Cyanosis – discoloration of the skin, nails, and various membranes; results from an increase in hemoglobin levels.

Cyanotoxin – a toxin that is produced by cyanobacteria; is commonly found in marine environments as a "blue-green algae" and causes the poisoning of Shellfish meat.

Cytotoxin – a toxin that is harmful to cells and the immune system.

Dracotoxin – a toxin produced by Weeverfish; can rupture and destroy red blood cells, and depolarizes membrane structures.

Devonian Era – time period dating back roughly 419 to 354 million years ago; was considered the "Age of Fish."

Dinoflagellates – groups of protists that are mostly made up of plankton and other various micro organisms; some are extremely harmful to humans if ingested.

Dorsal fin – fin or fins located on the upper part of the fish, helps with stabilization.

Echinoderms – a class of animals which consists of Sea Stars, Sea Urchins, Sea Cucumbers and Sea Lilies.

Enterotoxin – a toxin created by Rabbitfish; targets and kills cells in the intestines and abdomen which results in abdominal pain.

Euryhaline – an animal that can live in both fresh and salt water.

Filter feeder – a species that can strain food matter from the water into their mouths.

Fugu – a Japanese dish consisting of Pufferfish.

Gangrene – skin or muscle tissue that is dead or going through necrosis, typically caused by infection or loss of blood flow to a particular area.

Gastropod – a class of animal that consists of Snails and Slugs.

Heat labile – a protein that can be broken down or eradicated by applying heat.

Ion – an atom or molecule that has an electric charge by either losing or gaining an electron.

Maculotoxin – a type of neurotoxin that is created by the Blue-Ringed Octopus' (and possibly by the Flamboyant Cuttlefish's) salivary glands; this toxin is identical to

Tetrodotoxin and is considered one of the most potent venoms in the world.

Maitotoxin – an extremely potent toxin created by a certain species of Dinoflagellates; breaks down the membranes of cells and causes them to burst which can result in death.

Nematocysts – cells in the tentacles of Cnidarian species that contain venom.

Neurotoxin – a toxin that destroys nerve tissue; affects all nerve-related functions and tissue connected to the nervous system.

Operculum – an area that covers the gills or any aperture of a species.

Ostracitoxin – the former name for Pahutoxin; a toxin created by Boxfish that is highly poisonous and can discharge as red or green in coloration; has a detergent or foam composition.

Pahutoxin – formerly called Ostracitoxin; a toxin created by Boxfish that is highly poisonous and can discharge as a red or green coloration; has a detergent or foam composition.

Palytoxin – a toxin that is produced by certain types of Dinoflagellates which are usually hosted by a marine species (i.e.: Seaweed Coral); this toxin must be ingested or inhaled and can affect every living cell in a human body, typically leading to respiratory issues or death in many cases.

Pectoral fins – pair of fins located underneath the fish's head; helps control direction of movement.

Peditoxin – a type of protein toxin that is created by Sea Urchins; it typically enhances the effects of other toxins.

Pelvic fins – pair of fins on the underside of the fish by pelvis region; helps control direction.

Peptide – two or more amino acids linked together as a chain which will pair to other molecules in an amino group.

Polyps – colonial life forms that live on various marine life such as Coral or Anemones.

Polytheonamides – a type of polypeptide that is toxic; found in Sea Sponges.

Proboscis – an appendage that stems from the head of an animal; is usually a mouth or nose.

Saponin – a chemical compound that is discharged as soap-like foam; primarily created by marine invertebrates and some plants.

Saxitoxin – a type of neurotoxin that is created by a certain type of Dinoflagellate; it blocks ion channels and can be lethal if it enters the body; attributed to Shellfish poisoning.

Spicules – sharp pointed structures found in clusters on a species.

Symbiotic relationship – organisms that depend on each other for survival, or live with each other by choice.

Tetrodotoxin (TTX) – a type of neurotoxin that is extremely potent and blocks ion channels; found in Pufferfish, Porcupinefish, Ocean Sunfish, some Ribbon Worms, two Gastropods and possibly one species of Sea Star.

Trimethylamine N-Oxide – a metabolite found in some animals; it helps stabilize protein and counteracts urea but is poisonous to humans if ingested.

Zooxanthellae – a type of photosynthetic algae that lives in the tissue of other organisms; can also cause bioluminescence.

www.ingramcontent.com/pod-product-compliance
Lightning Source LLC
Chambersburg PA
CBHW070044210526
45170CB00012B/582